FRU GARDENING FOR BEGINNERS

THE BEST TECHNIQUES AND SECRETS TO GROWING YOUR OWN FRUITS IN THE HOME GARDEN

Bradley Gray

Legal & Disclaimer

The information contained in this book and its contents is not designed to replace or take the place of any form of medical or professional advice; and is not meant to replace the need for independent medical, financial, legal or other professional advice or services, as may be required. The content and information in this book has been provided for educational and entertainment purposes only.

The content and information contained in this book has been compiled from sources deemed reliable, and it is accurate to the best of the Author's knowledge, information and belief. However, the Author cannot guarantee its accuracy and validity and cannot be held liable for any errors and/or omissions. Further, changes are periodically made to this book as and when needed. Where appropriate and/or necessary, you must consult a professional (including but not limited to your doctor, attorney, financial advisor or such other professional advisor) before using any of the suggested remedies, techniques, or information in this book.

Upon using the contents and information contained in this book, you agree to hold harmless the Author from and against any damages, costs, and expenses, including any legal fees potentially resulting from the application of any of the information provided by this book. This disclaimer applies to any loss, damages or injury caused by the use and application, whether directly or indirectly, of any advice or information presented, whether for breach of contract, tort, negligence, personal injury, criminal intent, or under any other cause of action.

You agree to accept all risks of using the information presented inside this book.

You agree that by continuing to read this book, where appropriate and/or necessary, you shall consult a professional (including but not limited to your doctor, attorney, or financial advisor or such other advisor as needed) before using any of the suggested remedies, techniques, or information in this book.

Table of Contents

Introduction

Gardening can be rewarding in many ways. There are a lot of different approaches that result in different achievements. Before deciding how you would like to garden, you should understand what limitations you have and what you hope to get out of your gardening experience.

There are so many incentives for gardening. It might be for wholesome, fresh produce; for some peace of mind or stress release, or just because it is something your family has always done. If your doctor advised you to take up something that will keep you active and you're not into sports, take up gardening. You will never regret your decision.

There is never a wrong reason to garden. It is a hobby that gives back to everyone involved, including the environment.

Gardening can add to the beauty of your land. It can give your home that curb appeal you have always wanted. Adding color to a bland patch of grass can help to warm things to a new level.

It is highly beneficial to start gardening to lessen your grocery bill. A garden can give you fresh fruits, vegetables, and herbs for free. You will have a small amount of start-up money and some labor. You will also have fewer trips to the grocery store if you get a good crop growing.

Gardening can be very therapeutic for some people. The routine of doing the work can make them feel more energetic and disconnected from all the things that cause stress in our lives. There is also a gratifying feeling at seeing your results. Regardless of what those results are, you will be a success because you will have learned a lot from your experience. Things you will want to repeat and things you will not want to repeat. All are valuable lessons.

People have different intentions for gardening. They are often limited by their living arrangements. When you are ready to decide the style of gardening you are going to use, it is important to consider whether you want to arrange your garden indoors or outdoors. Outdoor locations might be limited due to sunshine exposure and drainage. The amount of space you must work with might help to decide your technique. The area you live in will help you to think about what you must work with. Your weather might cause problems or limit you to certain timing. It will also be helpful to recognize what you are trying to gain with gardening. You could just want an escape. It's possible you just want fresh produce. What a great reason!

All Methods of Gardening Share These Same Benefits

Two often unanticipated benefits of gardening are its delightful connection to nature and its remarkable stress-relief side effect.

Connection to Nature

Gardening is taking your place in nature. You are working with natural elements to bring forth delicious and nutritious produce to be eaten by your household. After you start gardening, you will feel more in touch with life outside your home. The insects, breeze, rain, and all other elements of Mother Nature will start to become more noticeable to you. You might be out at a friend's house and hear the thunder. Normally, you might think nothing of it. Since you have been gardening, you might think, "I hope my tomatoes can make it through a storm." Gardening will also allow you to escape the technology of the world. It allows you to stop and smell the roses. You can leave that cell phone in the house and take a short, electronic vacation. If you take the time and dedication, you can enjoy the smells, sounds, and colors of your garden. It generates a reason for you to get off the couch and go outside.

Stress Relief

People often garden for stress relief. It can be very therapeutic. The act of doing the responsibilities gardening requires can relax the mind. It is easy to get outside and enjoy the fresh air. The change in pace can allow your mind to clear and relieve you from the everyday issues we get stuck thinking about.

The pride in the success of a garden can make the work less of a chore and more enjoyable. The physical activity required releases endorphins, which in turn release a lot of tension built up from our everyday lives. You can be getting a large amount of exercise without even realizing it. You will also be getting sunlight, i.e., vitamin D, which is great for your health. It's a free vitamin D day when you garden. Soak up the sunshine and enjoy your new-found passion.

Chapter 1: Where to Start?

Start small. It's good to get excited and think about how you may develop many distinct sorts of products, however, you'll get overwhelmed if you tackle an excessive amount. In your first year, start small to enjoy and research what works for you, after which you may extend in the years to come. This is likewise a less high-priced manner to get started.

Plan What to Develop

When making plans for what to develop, it's crucial to pick out meals that your circle of relatives will eat as well as clean plants that don't require a lot of unique care. You'll additionally want to keep in mind your neighborhood conditions, consisting of the season, usual weather, and solar exposure. You can discover more about your neighborhood developing sector and usual first and closing frost dates by doing internet research of your region.

Figure Out the Nice Site

Pick a nice site on your lawn, whether it's a few bins, a raised mattress, a sunny region interior close to a window, or an area to position a vertical lawn. Be positive and keep in mind how much sunlight the site receives based on what you intend to develop. Additionally, you need your lawn to be positioned near a water source, your gardening tools, and your kitchen. To put together the site, you could take away grass, smooth up or flow vintage flowers, or tidy up your patio, deck, or a sunny spot interior to make room.

Decide on What You'll Develop In

In element 2, you'll discover ways to construct a few clean raised beds or planters. But in case you don't have the time or ability to do this, you could purchase a raised mattress package and, whether you're developing interior or outdoors, use character bins and planters. Even better, you could develop your flowers in recycled family bins in case you are in a decent price range or need to be more sustainable.

Build High-Satisfactory Soil

The high-satisfactory and fitness of the soil you are working on immediately influences how properly your flowers will do. When you're beginning out, it's crucial to shop for the proper soil and upload natural amendments to it over time. This is the premise for what we call "constructing your soil." Avoid the usage of lawn soil for bins; it does now no longer drain properly and is regularly too heavy for bins consisting of placing baskets. Instead, use potting soil, that's lighter. If you choose to go with hydroponic, you won't even want soil! How a lot less complicated can you get?

Plant Seedlings or Begin Seeds

For a beginner, shopping for seedlings at a nursery or lawn middle is the perfect manner to get started. However, in case you need to develop a specific type of veggie that your lawn keep doesn't have, you could additionally begin your flowers with the aid of using seeds. When the seedlings get large enough, you'll transplant them into your raised mattress or bins so that they can develop large and strong.

Establish A Weekly Gardening Schedule

To gain achievements with your city lawn, you'll want to know when you'll have time to work on it. This doesn't take all weekend. All you want are a couple of minutes right here and there all through the week to maintain matters, after which an hour or so at the weekend will help you do several longer tasks. Even when you have a hectic lifestyle, fundamentals consisting of watering, fertilizing, weeding, controlling pests, and pruning won't take up an excessive amount of your time. Gardening enables you and your circle of relatives to loosen up and gradually relax after a hectic day or week.

Deal with Troubles

You must address troubles on your lawn, and you also want the proper mindset to address them and discover solutions. This is the primary reason why a few novices quit. Plants may get sick, be broken due to neighborhood wildlife, war in the soil or conditions, or appeal to pests. The crucial issue isn't always to get discouraged. Even the maximum pro gardeners run into trouble. They have simply found out how to address them.

Harvest!

This is the praise for all your difficult work. The first time you choose that sparkling tomato or harvest a few lettuce leaves for a sandwich is awesome. You do want to test your veggies as they develop so that you don't end up with zucchini the size of a baseball bat! And when you have kids, that is the precise assignment to delegate to them—if you could maintain them from ingesting the entirety at the spot.

Understand Your Local Regulations

You might assume that developing meals at home is something that nobody has issues with. After all, you're doing an awesome thing, right?

Unfortunately, it's now no longer continually so clean. There are policies and rules you want to observe; otherwise, you chance dropping your lawn or being fined.

Every town or municipality has positive bylaws that manipulate what you're allowed to do for your property. These cowl fence height, accent buildings, smells and noises, and ugly premises. Before you begin your city lawn, make sure to study through the files that apply to you.

If you stay in a network with a homeowner's association (HOA), keep in mind that a few don't permit any meal gardens, even in the backyard. Check the policies and rules carefully. If your HOA does permit meal gardens, they'll be particular about the kinds of beds and bins you're allowed to apply, so your lawn doesn't detract from your neighbors' amusement or the aesthetics of the entire neighborhood.

If you rent, test together along with your landlord before you dig up the entire backyard! You can be constrained to the usage of a balcony, a deck, a patio, or an indoor area. If you do get a few areas in the backyard, keep in mind that someday you could want to transport and will lose your flowers.

Even if you observe all legal guidelines and rules, from time to time, the best thing to do is talk to your neighbors. Let them understand some of your plans for developing meals and cope with any of their questions and issues before you get started. Maybe they'll get the meal-developing bug, too!

Essential Tools for A Fruit Garden

Many of the things that you want are probably already around your home – especially if you're working on different outdoor jobs.

Here's a brief listing of a few helpful gardening equipment:

- Gloves can help you avoid hand blisters. Cotton gloves will be the most affordable, however, more luxurious gloves manufactured from sheep and goats last longer.

- A great straw hat with a wide brim can keep the sunshine off your face while letting in air to help keep you cool.

- A pocketknife or set of pruning shears is great for slicing side strings and blossoms.

- Sturdy rubber boots, lawn clogs, or boots repel water and deliver resources for digging.

- Bug repellent and sunscreen to keep you comfy and steady while working on the lawn.

Watering Hoses and Cans

Plants require water to grow, and when Mother Nature isn't cooperating, you may need to water frequently. For a large lawn, you would require fancy soaker hoses, sprinklers, and drip irrigation pipes. However, for plenty of small residence gardeners, a simple hose and watering can can work wonders.

Rubber hoses kink less compared to nylon or plastic pads; however, they may be a lot heavier to move around. Whatever fabric you choose, make sure to procure a hose that is long enough to reach flowers in each region of your lawn while not having to take water containers to reach more remote flowers. Decide on a hose that incorporates brass fittings and a washing machine included in the hose; some additions make the hose less probable to fail after extended usage.

Watering cans may be crafted from easy, cheap, brightly colored plastic or fancy metal. Vinyl is lighter; however, galvanized metal is rustproof and more appealing. Watering cans are available in many sizes, so it is best to know which size you need before purchasing. Ensure it is easy to cast off the sprinkler head for cleanup.

Hand Cultivators

A 3-pronged hand cultivator is a beneficial device to cut up clods of soil, straightforward seedbeds, and function in granular fertilizer. Additionally, after you plant your little box or extended mattress, a cultivator is a first-rate device to remove younger weeds as they germinate. When you're digging a planting hole, then a hand cultivator divides the floor easily in comparison to a hand trowel. Much like a hand trowel, make sure you choose a hand cultivator that feels cushy for your fingers as it incorporates a grip firmly fixed to the blade. The metal-bladed sort may last longer.

Spades and Shovels

Spades and shovels are simply part of the most-used gardening gears. A spade can be used for grinding as well as scooping and projecting. Shovels historically have curved and pointed blades, while spades have flat, right, almost rotating blades. A first-rate spade is crucial in any lawn for dispensing dust, manure, or compost. A spade is essential for trimming or breaking a sparkling floor. Many gardeners use spades for anything from slicing dust bags to hammering in bets. Very proper spades are rocky. Spades and shovels can either be long or short. A longer handle offers you greater leverage while digging holes, so keep this in mind in case you are shopping for a brand-new spade.

Garden Forks

Since a spade is for turning new lawn dust, I discovered that an iron fork is a much better device for turning beds that had previously been worked on. The fork sinks into the ground in as deep as 12 inches, it divides clods, and it loosens and aerates the soil more than a shovel. Iron forks look like short-handled spades aside from the 3 to 4 iron tines on their heads. The best ones may be those with 1 piece of metal with timber grips firmly connected. They are not only used for turning dust, but also for turning compost piles and root crops, like carrots and potatoes.

Garden Rakes

When you dig soil, you need to stage it, split clods of soil, and smooth the seedbeds (especially if you're creating raised beds). An iron rake is the best device for the process, though you can only use it in some instances. A 14-inch-diameter iron-toothed rake should have a lengthy, wooden cope that is securely connected to a metalhead. You might also additionally flip the metal head to ease a seedbed stage. Look for an aluminum rake to get a lightweight though a less-lasting model of an iron rake.

Buckets, Wagons, and (Additionally) Baskets

Though you don't own a 1,000-rectangular-foot lawn, you still need to move around seeds, fertilizer, gear, crates, and more. Listed under are three essential packing containers:

- **Buckets**: For potting soil, fertilizers, and hand gear, a 5-gallon plastic bucket is a first-rate box. You may obtain one from a construction site, so make sure you wash it out nicely. To get a greater, long-lasting albeit smaller bucket, then buy one constructed from galvanized metal.
- **Wagons**: For lighter items, like packets of seedlings, you can use something like a kid's old purple wagon. Wagons are first-rate for moving flowers and small bags of compost for your lawn. The wagon bed also allows you to preserve this stuff in the area – this is better than leaving it on the bumpy floor. If you're thinking about a wagon to move yourself (instead of gear) across the lawn, a brand-new innovation is using a chair as a saddle. This type of wagon usually has a swiveling chair and can be perched on four analog wheels, letting you take a seat down and push yourself which lets

you move at some point on the lawn. Its storage area is beneath the chair.

- **Baskets**: To acquire the best thing that will help you expand your harvest, invest in a cable or wicker basket. Wire baskets are less complicated to use as it is feasible to scrub the product whilst it's nevertheless from the basket. Wicker and wood baskets are great as they are aesthetically appealing and last even longer than those made of metal. Using a basket to pile in all your harvest is a lot better than trying to carry all the zucchini you have on your hands; it will also be easier for you to move them from your garden to your kitchen.

Wheelbarrows and Garden Carts

There will come a time where you need to move heavy things such as soil and fertilizer from one area to another for your garden or lawn. The best thing for moving stuff that is "bigger than a breadbox" are wheelbarrows and garden carts. The easiest way to distinguish the two cars is by using their wheels. Wheelbarrows have a single wheel together with an oval, alloy tray; garden carts have four wheels and a square wooden tray. Wheelbarrows are maneuverable in tight areas, can turn on a dime, and can dump things easily. A contractor-kind wheelbarrow has a deeper tray – this is worth the extra funding because of its first-rate quality. To get a light-weight wheelbarrow, look for one with a tray crafted from plastic. Garden carts are a lot higher, can move around larger things, and are less complicated to push than wheelbarrows. A large-sized garden cart can move around hundreds of soil, gravel, stone, and bales of hay without a problem. Some lawn carts have removable rear panels which make dumping less difficult.

Power Tillers

The traditional back- or front-tined energy tiller evolved to large-scale anglers which are now used by gardeners to turn their gardens during autumn and spring. The large power tillers (more than a 5-horsepower motor) are best when you have 1,000 rectangular feet or longer to till. Additionally, they may be essential for forming raised beds and dividing sod.

Chapter 2: What Is the Best Soil?

Add three parts of compost and one part of manure on any soil you have at your disposal. After that, rake the soil, break all the clumps, and ensure the surface is level.

Soil, sun, and water are the most important elements of gardening. Poor-quality soil can affect all the efforts you poured into your garden. Choosing first-rate soil is extremely important. What makes up a high-quality soil ideal for gardening?

Premium soil is well-aerated soil. This means that air has appropriately circulated through the soil giving it a healthy and bouncy feel. Soil that is too thick will not give the roots a proper chance to grow. Dense soils, like clay, will not make room for adequate water drainage as well.

Healthy soil should be free of other materials, like stones, that tend to obstruct growth. Sandy soil is a poor choice for a hearty garden. First-rate soil must be rich in organic matter, like ripened manure or compost.

You can make your compost by collecting peelings, pruning wastes, fallen flower heads, and even old tea bags and newspapers. These can all be turned into compost that's rich in nutrients. Just fill your compost bin with a mixture of brown and green materials and deposit it into your garden. Permacultures of worms and bacteria will grow into the bin. Mix the compost into your garden soil and the soil will nourish your plants.

There is also a different way of improving your soil through compost. Just deposit barks and leaf molds into the soil where worms and the weather can work their way in. Compost is the answer to soils that couldn't hold moisture well or to soils that are too thick. Fertilizers will, eventually, be unnecessary.

The ideal season to prepare your bed soil is in autumn. Prepare your plot by digging not more than 8 inches into the ground. Begin by loosening the soil. Remove all destructive materials like rocks and large woods. Start incorporating your organic compost into the soil.

Now all you have to do is wait for the next spring when planting starts.

The Best Potting Soil

If you're considering planting flowers for your lawn or developing your garden for flowers, you need to recognize which soil is exceptional for plant boom. The sort of soil you have on your lawn depends on where you're living, and it should simply be different from your neighbor's. This is because you should have exceptional soil to grow flowers. Additionally, you can know how great a certain type of soil is depending on the number of flowers growing on it.

Let's talk about outdoor and indoor soils so you can see the difference between them, and you can determine for yourself the exceptional ground for the boom of flowers.

Outdoor Soils

Outdoor soils are usually of 3 distinct types:

- Sand
- Clay
- Sand or silt

The best soil for flowers is one that is wealthy in vitamins, and this is sandy soil. However, don't feel bad if you think your soil does not meet the suitable situations; you could constantly alter its characteristics and turn it into top-quality soil with a touch of fertilizer.

Indoor Soil

If you've got flowers that are growing indoors, you may think that it will be a great idea to move the soil out onto your lawn and have your flowers grow there. However, that is a mistake and is additionally a horrific thought.

Garden soil incorporates microorganisms that could kill your houseplants. There are other alternatives in case you don't want to use the potting soil being sold in stores; you can sterilize the ground outdoors.

Sterilize the Outdoor Soil

If you choose to move your indoor flowers outside, you should first sterilize the soil to get rid of illnesses and remove bugs and weeds. After purifying the ground, you should alter the soil with peat and sand, so your flowers can have good drainage and the proper air so that there is ideal humidity.

You also can pick to shop for potting soil as they're very comparable and extra snug for your flowers. These soils encompass peat moss and vermiculite, which is a gradual launch of fertilizers appropriate for the boom of your flowers.

Both components will assist your flowers in developing robust roots due to the fact they'll be capable of reaping vitamins, humidity, and ventilation.

Know the Types of Soil for Your Garden

To acquire a dream lawn or have a well-stored nook of lush flowers, it's critical that you recognize the sorts of soil, their traits, and are aware of which one you want for every plant. The characteristics of the land are crucial!

Choosing the proper substrate is likewise critical for your plants to grow correctly. Each sort of soil has its specific bodily and chemical traits, and in the market, you could get numerous types depending on the specific needs of your plant.

A substrate could make your flowers more resilient against pests and different factors that could preclude their growth, besides defending them from parasitic illnesses.

Would you want to analyze what kind of land you've got for cultivation and which of them are more suitable for your plant type? Everything you're analyzing will be very beneficial.

- **Prepared Land**

It is the perfect substrate for pots because they have correct water drainage. In general, prepared soil is the only one used for the exceptional boom of decorative flowers in pots, planters, flower beds, and gardens.

Prepared land is most known for instantly growing a brand-new plant in a pot; this way, you could expect its top-quality from the beginning.

This sort of land contains additives, with the well-known ones being washed water, compost, moss, and rice husk.

- **Professional Land**

This sort of soil improves moisture retention and turns on herbal plant safety systems. It is a substrate that incorporates good-sized possibilities of nitrogen, an excellent detail for developing flowers. It additionally has an aggregate of black peat, peat moss, granules, and perlite. When is it beneficial? Professional substrates may be utilized in any crop, both indoors and outdoors.

Among the benefits of the use of this substrate is the plant could be healthier and more resistant to illnesses. Another benefit is that it is suitable for the basic improvement of flowers as the stems could grow to be more robust.

In summary, when you purchase professional land, your flowers can expect a higher metabolism and a low occurrence of illnesses. This fact will imply larger, stronger, and leafier plants.

- **Black Earth**

Black soil tends to be very uniform and is suggested for grasses and pots. Black earth is prominent precisely through its color (darkish black) and its origin (it becomes shaped after the decomposition of natural matter). Black soil is usually made from remnants of dry leaves, or the natural waste of animals that are absorbed as vitamins through the soil.

This sort of soil may be very beneficial while creating a crop or planting flowers. It is extensively used for the sowing of decorative flowers or additionally for the cultivation of food plants. Black earth performs an essential position as it incorporates vital vitamins in vegetables or fruits.

There are many benefits of black earth; however, we will point out its herbal potential to hold sufficient water for the improvement of the plant. Also, its additives are best for the circulation of the roots; in this way, the plant's growth is top-quality and wholesome.

- **Potting Soil Substrate**

It is good for foliage, vegetation, and veggies as it incorporates an excessive range of vitamins. The substrate is one of the primary sources for flowers as it provides fertility to the soil and vitamins to the plant. Thanks to the substrate, various issues regarding the growth of flowers may be avoided.

Among the sorts of substrates, we will especially highlight natural substrates and inert substrates. Among the natural substrates, one of the best is earthworm humus, a sort of substrate that gives more fertility to the soil due to its specific composition: nitrogen, phosphorus, potassium, calcium, magnesium, iron, and sodium, amongst different factors.

Furthermore, this substrate can inhibit the improvement of fungi and microorganisms that could affect flowers. Another of the exceptional recognized natural substrates is mulch, a sort of substrate this is liable for stopping early erosion, and still gives natural substances to the earth that help it battle against excessive or low temperatures.

Coconut fiber, perlite, gravel, and sand are in the category of inert substrates. This set of substrates has the perfect additives to nourish flowers. It also allows for the correct drainage of the soil and preserves the proper humidity for the flowers' improvement.

- **Leafy Ground**

Leafy soil improves soil texture and lets in a higher quantity of air. This type is taken into consideration as the "black gold" of flowers as it can offer a huge variety of vitamins (micronutrients and macronutrients).

Using it will result in your flowers being capable of germinating higher, their boom being a lot faster, and they will look stronger. By the time you decide to feature leaf soil for your vegetation, you may be routinely enhancing the composition of the ground, in addition to enriching it with several vitamins.

- **Organic Land**

Organic soil improves moisture retention and offers more resistance to the plant. A fantastic gain is that it can be used on all sorts of flowers. Organic soil is a part of the fertilizer that flowers must receive if you want to make their boom healthier.

Some professional gardeners use it as a lawn filler since this sort of soil can enhance the variety of vitamins in the land.

- **Special Lands**

Some special lands, because of their composition, are best in certain sorts of vegetation. This includes orchids. These soils can offer the vitamins that flowers need to appear beautiful, with green leaves and brightly colored plants.

Do you understand what sort of land to pick? Beautifying a lawn is more than directly planting a few flowers and watering them from time to time. If you need wholesome and robust decorative flowers, the soil is critical to acquiring that goal. The diet of a plant and the absorption of vitamins will depend on the quality of the soil you've purchased, so you'll want to shop for the proper substrate.

Having a lawn, land, flowers, and vegetation in first-rate situations isn't impossible; it starts through selecting the proper soil and with the appropriate composition. It gets a whole lot easier as you watch your little plant develop and grow stronger day by day.

Chapter 3: The "Lasagna" Method

The Dig Less Gardening, better known as DLG, isn't about getting rich off "Texas tea" from the ground. DLG is about becoming wealthy by learning about and then harnessing some of the true richness of the soil. DLG is all about becoming prosperous while working with nature rather than against it.

I've profited from the skills I've acquired over the past many years as a gardener. The food I've eaten, the nutrition, and the exercise are connected deeper with the root of this world, the true wealth of the living soil. Hopefully, this little DLG guide will provide others with some keys to unlocking and harnessing the power that exists in a healthy living earth.

DLG is a no-till or zero tillage operation over the life of a garden. DLG is about disturbing the soil as little as possible once the DLG is built. The goal is to do as little interference with the inhabitants of the soil by mimicking mature. In return, the rewards reaped include getting healthy plants for a healthier MBS and the health of the whole Earth.

Some people will ask "what's the point of Dig Less Gardening?" Hopefully, this book will help answer some of the many questions that can be raised by people that aren't familiar with a DLG approach to gardening. I think some of the answers to the question happen to be the many positive aspects when it comes to DLG, a form of no-dig or no-till sheet mulching style of gardening. One of the goals is about not disturbing the soil in the growing beds as little as possible so the life in the soil can do what it is meant to do.

People will often say they don't have a green thumb when it comes to gardening. Instead, they say they have a brown thumb. In the Dig Less Approach, a brown thumb is what is needed. This gardening technique is a twofold approach. First, there is the how-to construction part of a DLG garden. Secondly, and most importantly, is the brown thumb side working with the soil. If a gardener plays their cards right, the world that lives in the soil beneath us will do much - if not most - of the work for us. Once the DLG microbe palace is constructed, the microbe angels and saints of the underworld will work their magic.

DLG uses a simple technique that some might call lasagna or sheet mulching style vegetable gardening. That's where layers of organic materials are placed on a kill-mulch ground cover and then planted in. The ground cover used in DLG is cardboard.

Instead of digging up or tilling a lawn or other patch of Earth with vegetation like an established vegetable garden or some sort of patch of weeds, DLG might meet your needs. Often, using what is called a no-dig or no-till approach can be an awesome way to start a garden or move from a conventional dig and till gardening. In this Dig Less Gardening approach, gardeners can learn and then lean into this technique and turn away from digging and tilling the soil. Starting small and getting comfortable with the DLG approach is a good idea. Then with more familiarity, the use of DLG could be possibly expanded as the need arises.

Dig Less Gardening is intended as an introduction for the potential gardener that would like to learn how to grow vegetables without tilling or digging the soil. The techniques covered here are simple enough steps and inexpensive, but each step does take some time and energy as it includes gathering and installing the materials. Once the appropriate items, like cardboard, organic materials, compost, etc., are collected, the project can move forward rather rapidly. A DLG could be considered as a landscape installation project. That's because DLG is about building something lasting that can be productive year after year if attention is paid, to some basic construction and later maintenance practices after building the DLG.

A DLG is a sort of a permaculture-type garden arrangement. In other words, all the work that you put into the construction of the DLG pays off big time because you won't need to do as much in the future after the construction phase is completed. Once the DLG is put together and is up and running it's more like a perpetual growing machine. With some care, a DLG will provide healthy fruits and vegetables not only for many years but also contribute in a positive healthy fashion to larger issues the Earth faces like carbon sequestration.

After the DLG is put together and growing great food from the healthy creation, you'll hopefully find there is often less maintenance with things like weeding. That can be a winner for people that don't enjoy weeding.

Season one with a DLG is about construction, planting, and growing, and allowing the new garden to settle in with the Earth it rests on. Like a wine that can improve with age, a DLG should also slowly but surely develop and increase its health and wealth. In short, once the microbe temple is built, offerings are made at the altar in the form of carbon, the food of the microbe gods. Happy healthy Soil critters can equal happy healthy plants, healthy people, and a healthy Earth. Can I type that line too much? I think not!

This guide isn't just for novice gardeners. DLG offers a great deal to more advanced seasoned growers. DLG is a how-to lean into kicking the destructive conventional techniques and practices of scorched Earth gardening of tilling, digging, and nuking the Soil life. Hopefully, prospective readers will be comfortable and learn enough to try some of the Dig Less Gardening approaches.

Most of us only know about digging, turning, tilling the soil, and then planting. This Dig Less Gardening how-to guide is here to help introduce novices and others to the concept that tilling is not necessary, plus some of the multitude of benefits to the soil, plants, nutrition, and climate change. In many situations digging into the soil is no longer required to grow a vegetable garden. In some cases, people that own a rototiller will find they can eventually sell this dinosaur. After learning about and then reaping the rewards after trying out the Dig Less Gardening approach, the soil can change, the plants can change, we can change, and the Earth can be changed for the better simply by raising healthy vibrant soil and tasty awesome food.

Most people think they need to race off to a store of some sort and haul a bunch of fertilizer home to feed the plants they grow in their garden. That might be the case in traditional scorched Earth gardening but not so in DLG. With a little knowledge and a well-put-together DLG, feeding the garden carbon in the form of mulch and or compost should keep the microbes whistling while they work in the smorgasbord of carbon you provide and they get to chow down on. In a nutshell, the life in the soil eats the vitals that you provide as well as minerals in the soil, and they feed the plants and the plants feed the life soil all over again. They work together, bartering in a give and take sort of way. If we choose to work together with the soil-plant combo, we are all winners. The soil-plants, MBS, and Earth all gain in positive ways.

There is discussion today that many human illnesses, modern-day plagues, can be attributed to human's disconnection from the healthy microbes of the Earth. If this is turning out to be the case, and I believe it is, DLG provides an unparalleled health spa. Not only can a DLG provide nutritious food but also when we garden in healthy living Soil our body and microbes that make us what we are all happier campers. Happier campers can help create and maintain a happier world.

To wrap it up: Dig Less Gardening is a win-win gardening approach for novice and most other gardeners that would like to learn how to do sheet mulching, lasagna style, no-dig, no-till, fewer weeds gardening. Learn to create a traditional row layout garden where a lawn or other vegetation is or has been growing. Learn how to convert a lawn or other ground like a tilled garden into a Dig Less Gardening for fruits and vegetables. Plus, the big winner is of course health and wealth for the trilogy of the soil-plant, MBS, and Earth.

This K.I.S.S. (Keep It Simple Stupid) approach isn't limited to the creation of a square or rectangular vegetable row and mound garden. With some creative thinking and work, Dig Less Gardening can be adopted by gardeners to be a variety of shapes and for many different types of planting and garden styles and layouts.

This guide covers some things like, what is Dig Less Gardening, why Dig Less, benefits, equipment needed, how-to steps, tips and resources, and more.

Chapter 4: The Display of Your Garden

Deciding on which plant to grow will depend on your taste and suitability to the environment. Study the plants first so you won't regret what you planted. Stronger plants have greater immunity to pests and diseases. However, are they apt for your soil and location? As much as possible, always choose the varieties that are disease resistant and have strong soil compatibility.

Maybe you can plant something exotic that's not available in your farmer's market. Plant the vegetables and fruits that you find pricey or have poor quality in your local grocery store. Organic produce is especially expensive in the market. Grow them in your backyard and expect some major savings. Don't forget the freshness and tastiness that's always available in your kitchen.

You can start with packets of seeds or with seedlings that started in the greenhouse's nursery. Seedlings are more costly than seeds but it's a great choice for beginners. Seedlings will save you time. If you are choosing seeds, be sure to read the labels and instructions about watering and spacing. Labels will also tell you if you can directly sow the seeds into the soil or if it needs to start indoors. You can grow ample seedlings indoors while waiting to plant outdoors. Choose plants that appeal to your diet and be sure that they are suitable for your garden's size.

Types of Garden

Deciding where to grow your garden is an important task. This is where your plants will live and where you'll watch them thrive. The first consideration is always the sun. Where does the sunlight hit the most? That is where your garden will be. There are three types of garden you can choose from. Decide on one that will support your needs and lifestyle.

A traditional garden is a kind that will give you tons of options on plants that you can grow. The natural soil beds will help you utilize and maximize the nutrients found in the ecosystem. Though some insects and bacteria can help the growth of your plants, some bacteria varieties may be the deterrent factors. If pests and microbes develop into a problem, control them naturally. There are safe and biological controls available like microscopic worms which are effective in killing vine weevil grubs.

If you are an apartment dweller in the city, a traditional in-ground garden may be impossible. It's best to choose container gardening as an alternative. Several container types that vary in size, shape, and material are available in the market. There is no limit on the type of container you can use. The conventional clay pots and terra cotta are the most popular ones, but you can also employ recycled materials such as wooden barrels, old bottles, plastic containers, or galvanized metal buckets. Since space is the foreboding factor, you can try planter boxes on your windowsills or even hanging planters. Wall planting has become a revolutionary and ingenious technique in urban gardening.

When considering the container types, bear in mind their differences. Some can retain heat while some may lose moisture rapidly. Poke a few holes if you're aiming for DIY planters to ensure adequate drainage. Choose a container deep enough for the roots. If you are using organic potting soil from the store, container gardening is the best choice. Store-bought soil is already rich in nutrients, weed-free, and has aerated. Choose plants that are suitable for such a limited space and place them right where the sunlight is. Plants that are ideal for container gardening are spinach, lettuce, garlic, onions, green beans, carrots, and herbs. Tomatoes and peppers are also suitable, but they require staking for they will outgrow the container.

If you're looking for something that's between traditional gardening and container gardening, raised beds are your options. Raised-bed gardens let you control weed growth. It's also the perfect solution for gardeners who want to limit their physical movements and for those who are suffering from back problems.

Bricks, cinder blocks, rocks, and untreated wood are used to raise the garden beds. The height of the raised beds can range from 6 inches off the ground to two or three feet. Depth should be at least 16 inches. The width shouldn't be more than four feet to make sure that you can still reach the center. Raised-bed gardens are as good as traditional gardens. You can grow any plant using high-quality soil. Even deep-rooted vegetables like cabbages do well in raised beds.

Since the early days of man, plants have always been a natural part of our life. For many reasons, humans have always taken gardening as a choice activity, both for sustenance and for recreation. The beauty that comes with the natural gifts of having healthy greenery in our environment is another factor that endears man to gardening. Every time space permits, people are quick to start a garden in their backyards, the yard, or a field. From the strength that comes with the natural build of a fruit tree to the shade it provides, to the yearly expression of beautiful colors and the tranquility of a garden, all these factors contribute to creating a strong bond between man and his garden, not forgetting the savory taste of garden products.

The close links that man has managed to create with nature naturally are also a very important part of yards and gardens. Not only do we take our time to tend to our gardens, but we also pay keen attention to them to make sure they are healthy and well sanitized. With the right amount of space, you can convert your lawn to a pleasant haven of healthy vegetables, fruits, and herbs for either commercial use or domestic consumption. You may also incorporate a range of shrub plantings in your outdoor garden.

You may also have an outdoor garden in a limited space. Hanging baskets, container gardens, raised beds, etc., are ways of incorporating healthy plants into your garden to create a beautiful ambiance that gives life to the outer space of your home.

If you wish to venture into outdoor gardening, you must bear in mind that there are a lot of conditions and activities that come with this venture. You will need to deal with pests, weeds, and diseases. First, you need to get valuable knowledge on the best outdoor gardening practices. Get the most suitable crops for your location and opt for the healthiest varieties of plants. You also need to find the best outdoor vegetables, fruits, and herbs that will best contribute to helping you choose the best outdoor garden designs, or layouts but note that you are going to combine all these factors with dealing with the perks of outdoor gardening.

Outdoor Gardening

Physical and Mental Well-Being: Outdoor gardening is a very beneficial venture for one's overall health and maintains a healthy mind and body. Because of the activities or exercises that come with maintaining an outdoor garden, many people, especially seniors, turn to it as a recreational activity to cleanse the mind and keep the body fit. Outdoor Gardening involves a lot of physical activities. As you plant, pull out weeds, and dig, you get the chance to burn up to 400 calories every hour. This will not only help you to stay fit; your brain and mind will also stay sharp.

It has also been proven to be a great stress reliever, reduce blood pressure and cholesterol, as well as ease depression. According to studies, just looking at plants can result in changes in the heart, blood pressure, muscle tension, and electrical activities in the brain. When they play with colors and textures, gardeners can imbibe emotions of peace, tranquility, and joy. It is safe to say that gardening goes beyond mere recreation, as it also serves therapeutic purposes for the soul and body.

Environmental Benefits: These days, we tend to hear more about the effects of the activities of man on the environment, but little is often said about the great benefits of the green impacts of gardens to the earth. Just as it is with indoor gardens, the plants in outdoor gardens also serve as air purifiers as they take in carbon dioxide and other air pollutants and produce oxygen. Plants also produce fragrance. Other benefits of outdoor gardening to the environment come from such factors as plant covers like mulch, which keeps the soil firm. As a result, soil erosion is reduced, keeping sediment away from streams, drains, and roads.

When you plant new outdoor gardens like rain gardens, you will be making use of the run-offs from rainwater, such that harmful pollutants never make it to streams and other water bodies. Not only are traditional outdoor gardens like rain gardens good for the environment, but they also need less maintenance, and as such, they cost little funds. Let's not forget that these gardens are also very beautiful and pleasing to the eyes.

Beyond the benefits that outdoor gardens have for the environment, there are benefits of outdoor gardens that are important to man's well-being. It has been discovered that when you have a proper environmental landscape design, it will keep your home cool during summer and warm it up during winter. There are also energy-friendly landscapes which, according to the Plants for Clean Air Council, can help reduce energy costs up to 20%.

Increase the Value of Properties: You also get the chance to invest in the future by maintaining a beautiful outdoor garden landscape design. When your garden is designed beautifully, it makes your entire house more beautiful and appealing. This will also boost the whole value of your house. The monetary value that the garden can add to your house to help you sell your house faster than it would have been the case with a house without a garden.

Growing Green: Not to be cliché, but we always hear the phrase, "seeing is believing." If you are growing for commercial purposes, it will be easier for your potential customers to see what you are selling to them; the plants' freshness, and the quality of what they get from you. Rather than using artificial nutrients and light, as is the case with indoor garden practices, you can guarantee organic farming in your outdoor garden as the plants readily have everything they need available to them from nature. In this case, you are sure of what your herbs, fruits, and vegetables have touched, what they consume, and how they grow.

Self-Satisfaction: Outdoor gardening has gone beyond a mere activity for survival to become an art. Just as the diverse landscape styles that come with outdoor gardening can add beauty to your home, it is also a great satisfaction source. The land and space you use for your outdoor garden also afford you the chance to explore and try out a lot of new plants and styles. As you continue to practice, you will keep getting better at what you do, such that you begin to see new opportunities and exploit them as they come.

Types of Outdoor Gardens

When one speaks of an outdoor garden, the general notion may be the image of plants scattered across the place. The truth is that there's more to outdoor gardening than just putting seeds into the soil. Outdoor gardens come in many shapes and sizes, and they can comfortably accommodate different types of plants at the same time. Don't make the mistake of thinking of the outdoor garden as just flower or vegetable gardens. There is more to the outdoor garden. As you already know, plants are very good ways of added beauty and color to your home, both inside and outside the house. This is probably why more and more people continue to embrace gardening every passing year. It is important to correct the misconception that gardening is only about typical residential or commercial landscapes because, in truth, various types of gardening have to do with different styles, methods, locations, and plant varieties.

Outdoor gardens are generally divided into two broad types, namely, the ornamental and food gardens. The ornamental gardens are of different types according to the variety of plants grown in them and the methods used in arranging those plants. The most popular types of ornamental gardens are container gardens, rock gardens, wildflower gardens, formal flower gardens, informal flower gardens, and water gardens. A few of the things you will typically see in these gardens are shrubs, grasses, and trees.

Ornamental Gardens

Ornamental gardening is planting that involves growing different plants in one place for aesthetics. This system of gardening is designed to form a beautiful pattern for viewing pleasure. The first aim of this gardening is for aesthetics. The area where ornamental gardens are planted are designed to add beauty to the environment; hence gardeners that are into this system are not very concerned about crop production. In creating landscaping plans, ornamental plants are often incorporated into the designs, and as such, many states create ornamental gardens to attract tourists and for the viewing pleasure of those who live in such cities.

While this ancient practice is mostly used for flower gardens, the modern practice has found ways of incorporating food crops into this system, so ornamental gardens may also include vegetables, herbs, and fruits. It is natural for humans to want to create beautiful pieces for pleasure and relaxation, but it is also important to eat healthy foods. This is why gardeners have now found a way of planting food crops without leaving aesthetics behind. This system of gardening is simply the proverbial way of using a stone to kill two birds.

Gardeners now use food crops to form ornamental gardens around their homes, in public places like churches, galleries, or any other place dedicated to meditation. Imagine a beautiful garden where you can find orange fruits and savor the fresh scent of herbs. It is important to note that some religions have spiritual activities that are meant to take place in gardens, while other gardens of this type are meant to serve recreational purposes.

It is important to note that there's no standard layout for ornamental gardens. They may differ according to the types of plants grown and the environmental factors that surround the garden's location. While a few of these gardens may be highly majestic with an intricate arrangement of beds and lanes that are more appealing when sighted from afar, others like those meant to beautify small cottages are done in a very casual style.

The extent of ornamental gardens' beauty cannot be overemphasized, but beyond its beauty, one cannot neglect how useful these gardens can be. For ornamental gardens with fruits, herbs, and vegetables, one cannot ignore the fact that the products are very healthy, while typical ornamental gardens can serve as recreational spaces for sports, lounging, strolling, theater, or live musical performances. One may use such gardens for events like birthday parties, weddings, etc. as the garden's surroundings go a long way to add to the beautiful ambiance of the events.

Ornamental Fruits, Herbs, and Vegetables

When you opt for an ornamental vegetable, herb, and/or fruit garden, you get a brilliant burst of colors in your garden and your foods as well. One of the best reasons why ornamental edibles are becoming a favorite is that children can also get involved in the gardening process. They can join in the harvesting process, and they also get to eat healthy vegetables without resistance thanks to their vibrant colors. Studies have also shown that the colorful pigmentation that produces the colors has antioxidants that help guard humans against diseases.

Although the idea of maintaining edible ornamental gardens is not entirely new, it has continued to become more popular. In the past, we've had gardeners incorporate herbs into different parts of their outdoor flower beds. The truth remains that edible plants, regardless of where and how you choose to grow them, are never chosen as ornamentals before they are chosen as food. The first aim of growing herbs, vegetables, or fruits is to consume them as food. When they are used in ornamental gardens, they serve both purposes, and in that case, nothing is lost as many of the typical ornamental vegetables also act as decorative edibles. You may push to achieve both aesthetic pleasure and feeding pleasure with your garden without losing out in any aspect. All you need to do is to get proper knowledge of how this is done and invest quality time to get it done as well,

You will not be surprised if you find a gardener mixing both vegetables and herbs in the same container or garden bed. In an ornamental garden, this is often done by combining vegetable and herb plants with similar colors and textures to complement the rest of the garden's color scheme. To give a better boost to the beauty of your edible ornamental garden, you may choose to plant your crops in raised garden beds such that they are easily accessible. This will also give room for a better drainage system. Having a circular garden is another choice.

A wide range of vegetables may be added to the ornamental garden to give it an appealing look. These fruits and vegetables may be planted in containers or garden beds. If you plant appealing leafy fruits and vegetables in your garden, you will have a very interesting garden in terms of looks. The good news is that there are very many varieties of greens in this category to choose from. From the many shades of red hues, greens, purples, and bronzes to the fire red textures or the red sails; think about the loose-leaf lettuce that introduces the red-bronze color into the garden or the Cimarron lettuce, which invites more bronze to the garden. Rather than planting ordinary green romaine, you may opt for freckles as this variety is dotted with burgundy and immune to bolts, just like the darker variant of the burgundy Galactic, which has curled leaf edges. Rainbow chard is another beautiful plant that you may want to add to your ornamental garden as it comes in many colors. There is another form of chard known as the Bright Light. This type of chard comes with colorful stems and leaf veins of purple-red, orange, hot pink, red, and yellow. Thanks to its length, it also serves as a colorful backdrop for other plants.

If you are looking to add more colors to your garden, Carmen sweet peppers are a beautiful addition that could give your garden a burst of colors. This pepper comes in different colors, shapes, and sizes, so pepper lovers have a wide range to choose from. You may go for the greenish, purplish, whitish, reddish, yellowish, etc., variety as they are always readily available within a broad spectrum of colors.

To incorporate fruits into your ornamental garden, you may look towards popular vegetable fruits like eggplant. This plant also comes in many colors from greenish white, to dark purple, to pink or lavender. You may also find a few of these plants with stripped varieties.

One good way of adding vibrant colors to your garden's landscape is to plant tomatoes, with their bright cherry red fruit, which also has many varieties. You may get the purple, black, white, yellow, red, or the stripped variety. Another practical choice is the bean plant, which you may not know also comes in different colors. Try to get the purple or the yellowish-green beans to add more color to your garden. Remember the colorful bean blossom? This should be good for you. The ornamental scarlet runner bean blossom, which comes in a bright pink color, will give life to where it is planted.

The cabbage plant comes in different shapes and colors, just like cauliflower and broccoli. You may find the odd-looking orange cauliflower or the purple broccoli. As odd as they may look, they may be the plants that will invite your non-vegetable-eating family members to dinner. The perennials like the globe artichoke are plants that are likely to add diversity to your garden. With its beautiful leaves and attractive fruits that transform into hallucinogenic, blue-colored fruits that invite bees from a long distance, you definitely will want to stay back and stare at your garden for a long while. Try asparagus as well. With its lengthy, wispy fronds that look like ferns, you will succeed in creating a cool, beautiful edible garden. If you are in this for the long haul, rhubarb comes back to the garden each year by spreading its large leaves over the scarlet stalks that spring from the soil and are a viable choice too.

Another good news is that you can successfully grow plants like lavender as ornamentals and medicine. With its purplish-blue flowers, this plant grows in clusters and can serve many purposes for domestic and commercial use. Consider making essential oils, perfume, and indoor decoration from the product of your beautiful garden.

Food Garden

The food garden is the most common outdoor garden for growing herbs, fruits, and vegetables. Here, farmers use different techniques to grow, hone, and cultivate plants for domestic or commercial purposes. In many cases, fruit trees like apple or pear trees are used to enclose the garden space. In other cases, barbed wires or stakes are installed around the garden to form a fence around it. Many of these gardens are also left without barriers. It all depends on the gardener's choice. Within the barriers (inside the garden) are fruits, herbs, or vegetables planted in different styles on ground-level or raised beds. When planted on raised beds, the soil is given a chance to warm up during the early days of spring. The raised beds are also easier to keep. The crops in food gardens are planted in ways that are easy to maintain, even without synthetic pesticides. All that is required is the availability of organic matter and organic gardening techniques or other beneficial cultural methods like companion planting, crop rotation, or mulching. All these techniques are employed for the healthy growth or cultivation of food crops.

Other Outdoor Garden Types

Rock Garden

You may not have a large expanse of land with fertile soil on a flat plane to plant your crops. This may be a luxury in a few areas. Therefore there are other alternatives, like the rock garden, an option for a sloppy area. This is also a great alternative for locations where the water supply may not always be available. A rock garden is also a practical choice for people who desire to bring the desert's sparser beauty to their homes. Some plants do well when planted on rocks or gravels. Such plants create a unique beauty when planted in such conditions. Plants like the desert. Aloe vera plants do well when they are planted in such environments under warm temperatures. If you are resident in an area where the weather is mostly cold, you may go for plants that require cold temperatures like the ice plants, etc.

Water Garden

This is another outdoor garden that can bring the peace and tranquility associated with nature to one's home. This gardening involves a water body like a pond or a stream within one's location, and many plants grow and thrive by or in water. The peaceful ambiance that the presence of water bodies brings, can be both soothing and refreshing, and this can also endear certain animals to your space. You may then choose to match this serenity up with highly appealing plants by choosing from your variety of fruits, herbs, and vegetables. All you need to do is make sure that your chosen plant will survive in such a condition.

Community Gardens

Community gardens are public spaces given up to individuals to rent a portion of land where they can plant their herbs, fruits, and vegetables for domestic consumption or commercial purposes. These gardens allow people to get fresh farm products, even though they don't have lands of their own where they can grow crops. Ornamental gardens may also be planted in this space to beautify it and bring life to the neighborhood by giving residents a feeling of being connected to their environment. The care of these gardens is done collectively, and everyone who participates in taking care of the garden shares the products.

With such a garden, all individuals involved are given a sense of productivity as they get to share in the chain of food production. It is important to note that these gardens are not merely for growing vegetables, fruits, and herbs, as there is also room for planting native crops and medicinal herbs. There are also butterflies and ornamental gardens of this type. A community ornamental garden is often a public place for the display of the artistic beauty of plants.

Outdoor Gardening Methods/Layouts

Container Gardens

Container gardens open gardeners to an entirely different style of gardening, as opposed to the traditional gardening style. With this gardening style, farmers get the chance to move their plants from one spot to the other to get the desired nutrient. Container gardens are mostly used indoors, but they are also used to plant crops outside. This gardening method takes away the problem of dealing with weeds and other soil-borne plant diseases. Gardeners can also control the moisture level of their plants, and they can decide the level of sunlight and temperature that their plants are exposed to.

There is also the perfect chance to recycle with this method using items that are no longer useful in the home. Industrial materials that may have been abandoned or thrown into a landfill can also be used.

You may decide to use old boots, old buckets, or an old bathtub for this purpose.

Container plants may be placed individually in the outdoor garden or aligned in groups to add a beautiful look to the farm. These containers may be of different sizes, shapes, colors, and textures. The same goes for plants. The aim is to create a beautiful appeal on the farm using these containers.

Raised Bed Gardens

Just as is the case with container gardens, raised bed gardens help gardeners control the type, quantity, and quality of soil used in the garden. Because raised beds are often isolated structures made up of either concrete, wood, or stone, the soil underneath doesn't always affect the garden's outcome. With these structures, gardeners can plant medicinal, edible, and ornamental plants using unusual surfaces like rocks, hard clay, or even concrete surfaces. They give room for proper drainage and maintain the temperature of the soil. One factor that endears this system of gardening to people is the fact that it does not cost so much to maintain, unlike the traditional gardens.

According to the gardener's desires, the planks used in building raised beds are either screwed or nailed in different sizes and shapes. The standard practice is to make to have a relatively wide bed but making sure that the width is not too much. This is to help the gardener to reach the plants in the raised beds comfortably. The heights of the raised bed are typically about three to eight inches high, depending on the nature of plants grown in them. Good, healthy soil is used to fill raised beds. Compost and prepared manure are also mixed with the soil to give it a healthy boost for the plants. Elderly gardeners and the physically challenged use this system of gardening more because they can access it from a vantage point according to the bed's height. They can also sit while tending to their garden, so their joints and spine are not stressed or strained.

Tire Gardening

There is an appeal that comes with using recycled tires in the garden. The use of these tires also adds beauty to the idea of container gardening. To use tires as containers for your garden, simply get an old tire, and remove the outer rim to make it flexible. You may also work the tire's tread by pulling it out from the inside. This will give a visible, sleek look to it. The next step is to stack up treads, one after the other, to put the tires in perfect shape for planting. You may grow potatoes or other root plants in this container as it allows for enough space for the roots to spread. If your plant is the type that requires warm environments to grow, the black surface of the tire gives the best chance for warmth.

If you don't want the natural circular shape of tires, you can alter its shape by tweaking it into your desired shape. All you need to do is to get logs and fix them into the tire to give you the shape you want.

Rows

This is the most popular outdoor garden layout, which has a straight design achieved with long rows that span from the left to the right side of the garden. With the left, the right direction of the rows, the garden is exposed to adequate sunlight and perfect circulation of air. If the rows are directed in an east to west position, it becomes too shaded such that the plants grown in the other rows don't get enough sunlight and air.

When growing crops in rows, it is advised that you grow tall plants like corn or beans on the northern end so that they don't end up forming shade over small plants. On the other hand, taller plants like tomatoes are meant to be grown in the middle of the row, while the short crops like the carrot plants are to be situated in the southern part of the garden.

Square Foot Garden

The square foot garden structure is made up of grids that are 4x4 squares. The grids are characterized by either woods or strings fixed to the frame for dividing the bed into equal-sized sections. Each section houses one kind of vegetable, herb, or plant; the number of plants contained in each section must be calculated. This is done by dividing the least number of spacing inches required by 12. This will typically give you the size of a single plot. For example, if the closest space between your onion plants is usually about three inches, your calculation will automatically become 12 divides by 3. This will give you 4, the exact size of a plot. This means that four rows in a plot will have four plants, such that you have 16 onion plants all together in a plot.

Vertical

Just as it is with the indoor gardens, you can also grow vertical plants in your outdoor garden. This is a good layout for a person that does not have enough lawn space to practice every other layout or method. Instead of using the traditional system of having garden beds, you can use up the vertical space by planting crops along either trellis, by using hanging baskets, or hanging your plants upside down. You may also use stackable containers to help you grow plants in a particular part of your garden by placing pots on one another to create a structure like a tower. These are called planting towers, and they are used as vertical options for planting root crops like potatoes.

Chapter 5: Easier Fruit Plants to Grow

Growing Your Fruit

Growing your food is both satisfying and gratifying, not to mention a cost-effective way of putting healthy, eco-friendly food on your table.

There is a wide variety of plants to choose from. We will look at the selection of potentially viable fruits best suited for raised bed gardening. These delicious fruit-producing plants include strawberries, blueberries, raspberries, and blackberries.

Strawberries

Generally, all berries are susceptible to insect pests as well as hungry birds. However, if you take this into account when you structure a raised bed for strawberries, for example, you will ensure you have sufficient netting to keep them protected at ground level as well as from above.

Strawberries can be successfully planted through small holes in a sheet of gardening plastic, which is then laid in the raised bed and covered with soil. The plastic sheeting acts as a deterrent to insects that move upwards from the root system and attack the leaves and fruits of the plant.

Strawberries can grow in-ground beds, raised beds, and large containers. Light gauze netting is required to cover the entire raised bed to protect the fruit from flying insects and birds.

To produce that sweet taste, strawberries need to flourish in cold weather. You'll need a lot of sun and some row covers to protect the small plantings from extreme weather. Choose fertile soil without weeds. Place each plant around 18 inches from another. Strawberry patches can produce abundantly from three to four years. They are best harvested in springtime. Pick them in the morning without removing the short, green stub. Wash them in running water before eating or refrigerating.

Strawberries throw out runners that attach into the soil to begin a new plant. These runners can be cut off and replanted in new containers. Thus, a single strawberry plant is sufficient to start with.

Raspberries and Blackberries

Although raspberry and blackberry plants tend to become particularly invasive in a traditional garden, they grow particularly well within the confines of a raised bed where they are easier to control and give a prolific yield.

Bear in mind that if you choose to cultivate these particular berries, the plants will require adequate support for their vines.

Blueberries

Another delicious and much sought-after fruit is the blueberry that grows on a bush. The dwarf variety is perfect for a raised bed garden. Check the best variety for your needs and space (CPC, 2018).

Because blueberries do particularly well in more acidic soil, it is easier to cater to their needs in a smaller, separate raised garden bed. Regular checks on the soil pH will ensure a bumper crop of these tasty, succulent, and healthy fruits.

Cool, Succulent Melons

Certain melon varieties are great for raised beds. Among these are the cantaloupe and honeydew melons. Watermelons may also be a successful choice (CPC, 2018).

Remember, all types of melons need space, as their vines spread for many feet. If you have a sturdy support structure, the smaller varieties may be happy to climb. Melons also require heaps of water, so by keeping these plants well-hydrated, you will ensure an abundant crop.

Take care to manage the vines to avoid them moving in among all your other plants and becoming a tangled mess.

Crunchy Cucumbers

Lindsay Mattison (2019), in her upbeat and informative article, How to Grow Cucumbers: 8 All-Star Tips for Your Best Crop Yet, gives great advice on how to grow these juicy fruits.

Like tomatoes, cucumbers are usually grouped with vegetables. However, these botanically considered fruits are a wonderful asset to any raised bed garden.

These juicy, crunchy fruits thrive in well-ventilated soil, and they enjoy good drainage and the containment style of cultivation in a raised bed garden.

Cucumber cultivars are divided into two distinct types:

- Pickling cucumbers, which are often smaller with rougher, prickly skin, have a short life span.
- Slicing cucumbers are generally larger and juicier than their pickling cousins. They have smooth skin and a longer fruiting period.

You can successfully grow cucumbers from seed. These should be planted in well-draining trays, which will be kept indoors until after the last frost date when they can be transferred to your outdoor raised garden.

Mattison (2019) indicates that within two months of planting, your first cucumbers should be ready for harvesting. These wonderful plants will continue to produce fruit for as long as you keep picking ripe cucumbers off the vines.

Allow the cucumber to reach maturity but take care not to leave them on the vine too long, warns Mattison (2019). Cucumbers that grow too big can become soggy and yellow.

Cucumber vines require warmth and lots of good nutrients as well as compost. Their vines also need to be supported by a sturdy trellis. This will help lift the cucumbers off the soil, thus protecting them from crawling pests. Lifting the cucumbers also encourages better airflow and avoids the development of powdery mildew, says Mattison (2019).

Cucumbers make a great salad and are delicious when served as a chilled soup, which is ideal for hot summer lunches.

Tasty, Tangy Tomatoes

Although tomatoes are often classed as a vegetable, they are, in fact, fruits.

Most tomato varieties thrive in raised beds as they require lots of sunlight to produce prolific quantities of fruit.

Tomato plants grow well in rich, well-drained, acidic soil that has a pH of between 6.5 and 6.8.

To ensure your success in growing these fruits, be reminded that because tomatoes enjoy the warmth, seedlings should only be planted in raised garden beds when the temperature reaches a consistent 60 degrees F, suggests Lucy Mercer (2018) in her informative article, Top Tips for Growing the Best Tomatoes Ever.

Mercer (2018) gives a neat tip for ensuring your tomatoes develop into robust, prolific producers. When transplanting long-stemmed tomato seedlings lay the lower portion of the stem on the soil in the raised bed and cover it with soil. Gently bend the remaining portion upward and secure this area of the stem to a suitably sturdy plant supporter. Because tomato plants shoot roots from the nodes along their stems, these extra roots will secure the plant firmly in the ground.

Tomato plants require lots of good nutrition and water so don't spare the care! Turn in extra compost and keep these 'little gems' well hydrated.

To grow sturdy and upright, tomato plants should be well-supported. Without this, they not only flop into the raised bed, hindering the growth of their fellow plants, but untended tomato plants quickly get out of hand. Decide early on how you plan to support your tomato plants. A cage system may work, or perhaps you may prefer to use individual stakes.

You will also need to regularly cut back all the 'sucker stems,' advises Mercer (2018). This ensures the main plant can focus on fruit production, rather than wasting its resources on side stems.

Zesty Citrus

In her informative article, Growing Citrus in Planters, Cathy Cromwell (n.d.) states you do not have to live in the Sunbelt to grow your citrus. The wide variety of dwarf citrus plants that is available affords most gardeners the chance to cultivate these rewarding plants in their raised bed gardens.

Citrus is not only an attractive evergreen, but the exhilarating and fresh perfume from its delicate flowers will be an asset to your garden. The joy of picking your citrus fruit is a bonus to growing these amazing little plants.

According to Cromwell (n.d.), citrus requires a daily dose of a minimum of eight hours of sunshine. Full-spectrum grow-lights can supplement natural sunlight if your citrus is grown indoors.

Varieties of citrus that do well in raised bed gardens include mandarin, grapefruit, lemon, and orange. Cromwell (n.d.) further indicates these dwarf citrus varieties are susceptible to the cold and should, therefore, be brought indoors during the winter months, where the climate is very cold.

In milder climates, plants can be well-covered with a suitable frost protection cloth such as a Garden-Quilt Cover.

Kumquat and lime are the most resilient citrus capable of withstanding temperatures between 20°F and 32°F.

The Correct Soil Mix

A good-quality inorganic planting mix, specially formulated for planters, will be best for your citrus. The percentage of vermiculite is important, as this will allow for good drainage as well as root aeration.

Mix some larger particles, such as wood or pine chips, into the vermiculite to create more air pockets around the roots.

Fine soil and compost are not recommended for citrus as they tend to compact around the roots, thus stifling them.

The Secret to Successful Planting of Citrus

Cromwell (n.d.) says it is important to locate the 'graft union' on the trunk. Remove any shoots below this point and make sure the 'graft union' remains above soil level when you transplant the citrus plant into your raised bed garden.

Feeding and Fertilizing Citrus

The roots of all containerized plants are forced to remain within the confines of their pot. For this reason, it is essential to offer sufficient water and nutrients to the plant to ensure its continued growth.

However, overwatering will cause root rot. So, a good rule of thumb is to water sufficiently to ensure the water drains out of the bottom of the pot. You should establish a watering routine where you use a moisture soil meter to gauge the level of hydration around the roots. Only water the citrus plant when required.

You may find when the weather is particularly warm; your citrus will require water more frequently.

Citrus are generally 'hungry' plants, requiring regular added feeding. Cromwell (n.d.) suggests a good fertilizer that contains phosphorus, potassium, and nitrogen as well as trace elements such as iron, manganese, zinc, and magnesium.

Blackberries

Blackberries are the type of plants that can bear after your first year of planting. They can continue bearing more for several years even with minimal care. Some types of blackberries are semi-trailing and would need trellis or fences to maintain the upright position. Guide and prune them regularly like you would a rose. You can plant them in either a sunny spot or a semi-shaded spot with good drainage soil. Include a good 3 inches of compost for a year. Planting is best done in early spring or even late winter. In the middle of summer, prune the top not more than 6 inches to trigger growth. Forty days after it has bloomed, your blackberries are ready for harvest. Careful harvesting should be done in the morning while they are still cool. Refrigerate immediately to avoid spoiling.

Cucumbers

Cucumbers are the best choice for those delicious pickles. These plants can be very rewarding to new gardeners for they mature rapidly and are extremely productive. They thrive best in warm weather. Cucumbers can be sown directly into the beds a couple of weeks after the frost of the last spring. When the warm temperature is constant and the soil has enough moisture, cucumber seeds can sprout within just five days. Plant them in the sunniest area with a spacing of about 6 feet. Mix your compost 2 inches deep into the surrounding soil. Harvest them as soon as you see your ideal fruit size. Pick them regularly because cucumbers tend to double their size in a day. Snip them, scrub, pat, and refrigerate to preserve freshness.

Eggplant

Eggplants will flourish during the long summer seasons. You can begin by growing the seeds indoors six weeks before spring. Keep them under fluorescent light for 16 hours while winter is still howling outside. Then transplant them to containers until leaves start to appear. When they reach 8 weeks, you can plant them directly to your soil beds. One plant can produce up to 5 pounds of eggplants in over two months. Loosen the beds before planting them in a sunny spot. Mix in your compost and give them enough water. Provide a 2 feet spacing for each eggplant and prepare some stake for support. When they start bearing fruits, press a firm thumb on the smooth, glossy skin. If it quickly bounces back, then it's harvest time. Prune them while keeping the cap intact. You can store them in a cool room even without refrigeration but no more than a couple of days.

Grapes

Grapes can grow in almost any weather condition but to produce that sweet taste, it needs to thrive in a colder climate. Grape seeds can start indoors four to six weeks before spring. Plant them in the beds in early summer with full sun exposure. Make sure that you dig the soil deep for the water drainage and place about 2 inches of compost. Plant them with a 7 t0 9 feet spacing and place trellis for support. Prune grapes regularly to prevent the growth of the disease. When fruits start maturing from red or white to black or blue, they are ripe and ready for harvest. Snip the cluster and don't wash them. Refrigerate in closed plastic bags.

Melons

Mouthwatering and packed with nutrients, melons are best grown in a warm climate. In humid weather, melons can mature in just 80 to 100 days. They need fertile soil with good drainage and 3 feet spacing. Mix in the 2-inch compost and let them have enough water. You'll know they're ready for harvest because melons tend to slip from their vine when they are ripe. You can easily tug the fruits from the vines. Keep melons at room temperature for 3 hours and not more. Then move them to a refrigerator. This will help bring out all those juicy flavors.

Peppers

Bring some heat into your garden! This striking fruit grows best during the warm summers. Plant these spicy seeds indoors for a week before moving them outside. Pick the spot where the sun is brightest and plant them in a fertile, aerated, and loosen the soil. Combine an inch of compost with the dirt. Water them generously. You can harvest mature peppers in different colors- green, yellow, red, or orange. Snip them from the plant without removing the stub and put them in a freezer.

Goose Berries

Before you consider growing Gooseberries, check and make sure they are legal in your state. Goose Berries are commonly afflicted with White Pine Blister Rust and are forbidden in states where there are many White Pine trees.

If you are permitted to grow one, make sure you have the proper amount of space, these bushes can reach heights of five feet and be just as wide!

Gooseberries are popular in jams, jellies, and preserves. Gooseberries are also used to make tea, as a flavoring for soft drinks, and sometimes made into wine.

There are several different varieties available and all of them taste different, so be sure to do some research first, to find one that you like best.

Plant these in a high sunlight area and use well-drained soil. You should be able to harvest some berries after the second year but waiting until the fourth or fifth year will tend to produce a better-tasting berry.

Gooseberries are a good source of Pantothenic Acid, Vitamin C, and fiber. By the way- no one is sure why they are called "Goose Berries." Geese don't like them.

Apples

An apple tree is a fine addition to just about any backyard able to accommodate one. The best thing to do, if you're interested in your apple tree is to visit a local orchard to see which varieties do well in your area. You may also want to purchase a sapling, since growing a tree from seed will take more time than most would want to wait.

Apples are not native to North America but were brought here in the 1600s by European settlers and the first apple orchard was planted in 1625. Apples have a long history of human use and may have even been the first tree to be cultivated by people thousands and thousands of years ago.

Most people eat apples raw, but they are also good baked and are commonly used in pies and other desserts as well as juices, ciders, and vinegar. Culinary uses for apples are nearly limitless, making them a good choice for a backyard garden.

Apples are a healthy alternative to common snack foods and pack quite a nutritional punch. High in fibers and a decent source of trace minerals, the apple is a solid choice for your garden.

Aside from the fruits and berries mentioned above, the perennial family also contains many useful herbs that are also simple to grow and care for.

Figs

SEED OR PLANT: is planted in autumn or early spring.

CLIMATIC REQUIREMENTS: should be placed in a sunny position and perhaps sheltered by a wall, because it is not very resistant to the cold and can remain outdoors without protection only if the winter temperature is not particularly cold. It loves dry climates and cannot stand humid and foggy areas.

CULTIVATION: the soil must not be too compact but light, well-drained, and slightly calcareous. Irrigation must be regular and administered in such a way that the soil of the pots does not dry out completely.

HARVEST AND USE: the re-flowering varieties ripen in June and August, the others from mid-summer to early autumn depending on the variety. The figs are ready for harvest when they become soft, and their skin cracks.

Citrus

SEED OR PLANT: the citrus plants obtained from seed bear bad fruit; therefore, it is preferable to buy ready and, above all, already grafted, plants, to be transplanted in late spring when the danger of frosts will have passed. The container for an adult plant must measure at least 18 inches in width and depth.

CLIMATIC REQUIREMENTS: citrus fruits prefer warm climate zones and mid-shade exposure, sheltered from strong winds but not light breezes.

Their ideal temperature is between 55 and 86°F (13 and 30°C), therefore in very cold regions, strong winter protection must be provided. The pots can be placed in an unheated but bright room, or, but only if the frosts are not too long, you can wrap the foliage of the plants with a cloth of TNT fabric and wrap the pot with insulating material.

CULTIVATION: They grow well in light, fertile, and rich in organic matter soils (special citrus fruit soils are available on the market). They want abundant and regular watering with non-calcareous water, which in winter will have to be reduced but not completely suspended: citrus fruits do not withstand prolonged drought. However, they fear water stagnation: to avoid it, keep the pots raised from the floor, resting them on a support. Every year, at the beginning of spring, good fertilization with a complete and iron-rich fertilizer is indispensable.

HARVEST AND USE: Citrus Fruits bear fruits from November to March, but some lemon varieties bloom and bear fruits throughout most of the year. The fruits can remain on the plant long enough, even when they are completely ripe. Harvest Citrus Fruits by cutting them off with pruning shears or by pulling them stalk from the tree.

CHOICE GUIDE: citrus fruits are at the top of the list of plants capable of satisfying both the palate and the eyes. Beautiful and decorative for the shiny, evergreen leaves, the intensely scented flowers, and the yellow or orange fruits.

Due to their limited development, the most suitable for cultivation in pots is the Mandarin (Citrus Reticulata), the Calamondino (C. Mitis), a hybrid between Mandarin and Kumquat, and the Lemon (C. Limon) especially in the "Florentia" variety which flowers and bears fruit in every season.

Peach

SEED OR PLANT: the bare-root plants are to be buried in winter, during the vegetative rest period. Potted plants, on the other hand, can be transplanted in February.

CLIMATIC REQUIREMENTS: to be at ease, the Peachtree must live in areas with a fairly harsh winter climate.

CULTIVATION: wants a medium texture soil, fresh but free of water stagnation, while it does not tolerate calcareous soils. Irrigation must be regular during the hot season, and fertilization, to be administered at the end of winter, must be rich in nitrogen and potassium.

HARVEST AND USE: the harvest period is from June to the end of August for the common peaches and from July to August for the nectarines.

CHOICE GUIDE: Only varieties specially selected for the pots can be grown on the terraces.

Apricot

SEED OR PLANT: it is planted during the winter (but not on the coldest days), taking into account that, once adult, the plant will need a pot 18 inches wide and deep.

CLIMATIC REQUIREMENTS: the apricot blossoms quite early, and therefore can be exposed to late frosts. It prefers mild climates, and in any case, where it is protected from cold returns and northern winds. It grows well in dry and little rainy areas, where the risks of cryptogamic diseases are also lesser.

CULTIVATION: it prefers light, warm and permeable soils because its roots cannot stand water stagnation. It should be fertilized at the beginning of the vegetative recovery with a slow-release fertilizer.

HARVEST AND USE: the fruits ripen from the end of May to the end of July, depending on the variety and climate zone.

Chapter 6: Some Common Mistakes with First Time Growers of Fruit Trees

Container gardens tend to have fewer problems than normal gardens because you use clean soil, and it is a closed system. However, you can still encounter issues is to help you avoid some of the common mistakes people make when starting their container garden.

Fill Large Containers in Place

Large containers are very heavy when they are filled with soil so you need to move the empty container to its final position and then fill it. This way you are going to avoid damaging yourself.

Make Large Containers Lighter

If you are planting large containers with shallow-rooted plants then you can fill the bottom third or so of the container with filler material, e.g., empty plastic soda bottles, and then cover it with some plastic screening. This helps make the container lighter as well as reduces the amount of soil required to fill it.

Avoid Overwatering

One of the most common problems people have with a container garden is drainage. All your containers need to have plenty of drainage holes and free-draining soil unless the plant specifically has other requirements. Be careful of using saucers under your containers as these collect water and will need emptying to avoid your plants getting waterlogged. Remember, test the soil with your finger (stick it in up to the second knuckle and see if the soil is dry at your fingertip) before you water. Overwatering can cause the leaves to yellow and drop, also a sign of mineral deficiency, but check the soil isn't too wet before you apply extra fertilizer. Check your plants after heavy rain showers and pour out any excess water pooled in your containers.

Avoid Underwatering

This goes without saying really and you are walking a fine line between watering too much and too little. You will usually need to water once a day as a minimum though on hotter days expect to water twice a day. In hotter climates, an irrigation system can save you a lot of work watering! Soak your plants when you water them until water comes out of the bottom of your pot so that all of the soil is soaked, and the roots can get to the water.

Rescue Dried Out Pots

If your container dries out then put the entire container in a bucket of water and leave it for about 20 to 30 minutes (don't submerge the plant, just the container). This should give the soil enough soaking that the plant can be rescued. For larger pots (which won't fit in a bucket) poke holes in the soil (carefully) with a skewer or stick and then water generously. Wait for about an hour and then water again.

Plant to Pot Ratio

Always consider how your plants look in your container when planting. A large container with lots of small plants will look peculiar and "not right" because the size of the container makes the plants look stunted. If you are planting for aesthetics then you want at least one plant per container that is as tall as your container and to have plants that tumble over the sides to help mask the harsh lines of the pot.

Avoid Damaged, Sickly or Weak Plants

The temptation to get a bargain at your garden store and buy the reduced plants is high and sometimes you can resuscitate them. However, check any plants you buy and make sure they are not damaged or diseased and are in generally good health. The bargain plants can be great, but you can end up with a high mortality rate with them. Damaged plants should be avoided as the damage can often introduce disease into the plant and therefore into your container garden.

Don't Be Scared to Prune

A lot of people are frightened of pruning their plants because they don't want to damage them. Most vegetable plants will not require pruning, but some herbs (lavender and rosemary) and fruit trees will. Other flowers and shrubs can also require regular pruning to keep them looking good. You can save some money by potting on prunes from shrubs using some rooting powder and growing new plants from these cuttings.

Ensure Good Neighbors

Make sure that you group plants with similar requirements for soil, soil, water, and food together in the same containers. Grouping plants together that have different requirements will end up with some dying. You can use double potting to put plants with different requirements in the same container. The plants are potted up in individual containers and then these are put into a larger container.

Feed Your Plants

Remember that your plants require regular feeding as the nutrients in the containers will soon be used up. During winter, you don't need to feed most plants but during spring and the growing season, they will need regular feeding. How often and how much they need feeding will depend on the plant so check out the labels to know how to treat them. A handful of pelleted chicken manure applied in the winter to each pot will help restore much-needed micro-nutrients to the soil.

Don't Be Afraid to Discard a Plant

Plants die. Sometimes from disease, sometimes from the weather and sometimes they just die. Do not be afraid to get rid of a plant if it is looking dreadful, particularly if your container garden is meant to look good. If the plant is diseased then you need to get rid of it as soon as you can to prevent the disease from spreading.

Be Realistic and Plan

Think about your lifestyle when setting up your container garden and plan so that your container plants will thrive. If you travel a lot, particularly in summer then you need to install an irrigation system or use self-watering containers or make good friends with your neighbors!

Grow What You Like

Your container garden is there for you to enjoy so plant what you like in them! If you like tomatoes, grow tomatoes, but if you don't, then don't grow them! Don't feel you have to grow things just because you ought to. If you like flowers then grow those. You can even grow vegetables and plant flowers in the same containers so not only do you get the color from the flowers but the produce from the vegetables. Your container garden is there for you to enjoy therefore you need to plant it with the things you want to grow because you will then be more motivated to look after and care for it.

Don't Put Gravel in the Bottom of a Pot

It is a myth that gravel, pot shards, or stones in the bottom of a container help drainage. It doesn't. These can be useful to ensure the soil doesn't block the drainage holes, but they do nothing to help water drain from your pots and certainly cannot be used instead of drainage holes. All that happens is the roots grow into the gravel in which water pools, so your plants still get waterlogged!

Check Out the Light

Plants need light to grow so you need to be aware of where the sun is in your garden and how it tracks across your containers. Position the sun-loving plants where they get maximum sun and those that like partial shade can go in the areas that get less sun. Remember that plants cast shadows that you can use to your advantage in shading tender plants, but the same shadows can prevent plants from getting enough sun.

Make a List Before You Buy

A shopping list for your container garden is essential when starting. If you don't then you end up buying stuff you don't need and not buying stuff you need. You end up with far too many plants and not enough pots and similar problems. Always make a list of everything you need and then stick to it (easier said than done) when shopping for your container garden to ensure you get everything you need.

Keep Plant Labels

Plant labels contain useful information such as how big your plant will grow, what light it needs, how much food it needs, and more. A lot of gardeners will throw these away not realizing that they can be very useful later when you are either trying to remember what a particular plant is or trying to work out what to plant with it. Keep the labels safe so that you can refer to them when you need to and give your plants the care they need.

Acclimate Your Plants

Plants can take a fair bit of abuse, but they generally don't like sudden changes in weather or light. If you move seedlings outdoors without giving them a chance to get used to it you can end up either with a lot of plants dying or their growth is slowed as they adjust.

Young plants or plants which have spent a lot of time indoors need to be slowly introduced to the outdoors over a couple of weeks. Put them out during the day and bring them in at night and repeat, leaving them out longer into the evening until they are out all night. Remember though if there is any risk of frost to bring them back in again. The process is called hardening off and is vital if you grow from seed.

Troubleshooting for Most Common Problems

No garden is perfect. Like life, you are going to experience problems in container gardening. You will make some mistakes along the way, but do not let this stop you from enjoying your garden. You will probably ask yourself what you did wrong that your vegetable plants have dry leaves, or brittle stems even when you were very careful in watering or fertilizing the plants. Do not worry. The chances are that these problems were caused by outside factors such as unpredictable weather, and/or temperature.

Listed below are common problems and concerns in container gardening:

1. My plants are wilting, even when watered frequently. What should I do?
    - There is a lack of oxygen due to insufficient drainage and aeration.
    - Too much water because drainage problems block the oxygen's access to the roots.
2. Check the drainage holes of your container, and unblock if clogged, or increase the number of drainage holes.
    - Use a growth medium that is porous, or more lightweight.
3. My plants look small, purple, and stopped growing. What happened?
    - Conditions were too cold.
    - They are not receiving enough sunlight.
    - Relocate the plants to an area with ample sunlight, and a warmer temperature. Chances are, cold temperature or cool conditions do not work well with your plants, thus, preventing the plant from growing because it cannot take up phosphorus from the soil.
4. The leaves and branches of a healthy plant droop once they are watered. Why did this happen?

- They have an insufficient drainage structure.
- Check the drainage holes of the container. Excess water blocks oxygen from entering the roots and soil.

5. The leaves or fruits of my plants have holes and irregular shapes. How do I fix this?
    - They are damaged by insects and pests such as leaf miners, worms, or bugs.
    - Remove the insects and pests by using insecticides. Be careful in choosing the type of insecticide that you will use.
6. I noticed that some leaves have white mold on them. What does this mean?
    - They have a fungal disease.
    - The weather is too cold/damp.
    - This is a fungal disease called 'Powdery Mildew' that is common to plants affected by cold, or damp weather. Relocate the plants to an area with ample sunlight and a warmer temperature. Remove and discard the affected leaves.

Aside from the solutions to these common gardening problems, here are some additional tips:

1. Clean the containers thoroughly before using them. This will kill any pests and remove leftover fertilizer salts.

2. Some plants are very sensitive to cold weather, requiring protection from the wind and the cold such as mulching, windscreens, or relocating to a warmer area.

3. When planting vining plants, secure them to stakes, poles, cages, or other support structures to prevent the plants from bending, or blowing over.

4. Elevate your containers using bricks, or wood blocks, and place a saucer, or a small pot to catch the drained water. Change and clean regularly to prevent mosquitoes from breeding in the water.

Chapter 7: Useful Tips

Determine Your Growing Zone

Knowing your growing zone will help you understand your general growing climate and which plants will likely thrive there. Divided into 11 categories, growing zones are calculated based on the average annual minimum temperature in each location.

Growing zones were created to help gardeners understand which perennial plants will survive the winter in a particular location. Because annual plants complete their life cycle in one season and do not survive the winter, growing zones technically do not apply to them.

Although your zone won't matter for most vegetables and herbs you grow, knowing it can help you understand which plants grow best in certain seasons, which plant varieties may grow better, and which plants may not grow at all.

For example, gardeners in cooler zones 1 to 4 might grow cool-weather crops all summer, but their season isn't long enough to support heat-loving crops such as okra and large melons. Gardeners in warmer zones 7 to 11 usually can't grow cool-weather crops such as lettuce in the summer, but they can grow them in fall and winter.

Seed packets, seed catalogs, and plant tags are great resources. They specify if a plant can grow in your zone.

Identifying Microclimates

In addition to knowing your growing zone, it's important to consider factors unique to your yard. The nuances of your specific location, or microclimate, alter your growing conditions and can also affect when you get frosts. Frost dates are key to knowing when and what to plant.

Microclimates can occur around structures, mountains, bodies of water, slopes, and more. Gardens planted next to a south-facing wall, for example, may have higher heat that prevents tomato fruit from forming in warmer zones, but creates a greater yield of peppers in cooler zones.

Gardens in the shadow of a mountain, in a valley, or at the bottom of a slope may get frosts later in the spring and earlier in the fall than surrounding higher elevations. This type of microclimate may require a gardener to delay summer planting or to cover frost-prone plants.

These are just a few examples of microclimate variations, and you should pay attention to how your plants respond to your microclimate. Make a note of your area's average last- and first frost dates and compare them to what you experience from year to year. If your conditions vary from what is considered average for your area, you can adjust your crops and the timing of your plantings accordingly.

Consider Your Garden Type and Space

Most plants are happy to grow in containers, raised beds, or the ground. However, some are better suited to one form or another.

Some herbs (e.g., mint, lemon balm, and oregano) spread aggressively, which makes them ideal for a container. If you plant them in a raised or in-ground bed, they will take over and smother other plants.

Some crops are considered "heavy feeders," which means they need more nutrients than others. Therefore, although heavy feeders such as tomatoes, squash, cabbage, and broccoli grow well in containers, they enjoy the moisture-holding capacity of raised beds, and the greater access to soil nutrients reduces the need for regular supplemental fertilizer.

Raised beds are also ideal for crops that require more plants for a plentiful harvest, such as beans, peas, and okra.

Some plants also need more space than others. If you're looking to maximize the yield of your garden, a container full of greens that can feed you for months may be a better choice than a container with one cabbage plant that will give you one head. Pick Plants That Get Along

Vegetables and herbs benefit from companion planting. Scientific research on the topic remains scarce, but we do know that the more diverse your vegetables, herbs, and flowers are, the healthier your garden will be. Some plants do seem to offer a measure of pest control to nearby plants. For example, icicle radishes may repel squash bugs and cucumber beetles. Conversely, some plants negatively affect others when planted nearby. Broccoli and tomatoes, for example, both high uptake levels of calcium. Planting them together could result in smaller broccoli heads and blossom-end rot in tomatoes.

Also, beware that plants in the same families (e.g., cabbage and cauliflower) often are visited by the same pests, are afflicted by the same diseases, and take up similar nutrients from the soil. (Plants that share similar characteristics, such as appearance, general growth habits, and seed and seedling presentation, are grouped into plant families.) In addition to pest control and nutrient needs, there are other benefits to companion planting. Vegetables with varying growth habits and nutrient needs benefit each other (such as planting cooler-weather crops in the shade of heat-loving ones). This is also an excellent way to maximize a small space.

Although some companion planting combinations may benefit your garden, don't dive too deeply into it now. Start with these suggestions and experiment with other combinations later.

Garden Maintenance

Watering as needed. The following guidelines are just that—guidelines. It all comes down to a matter of common sense and paying attention to your plants. If the ground is hot and dry, the plants are going to look parched. Don't leave them like this for very long. Going from one extreme to another is hard on a plant. It stresses their system and affects their ability to perform at maximum capacity.

- Plants that are in the ground usually require the least amount of watering. Unless you are experiencing a drought or excessive heat, most flowers will do just fine with watering two or three times a week. Again, this depends on the amount of rainfall you have, the temperature, and the type of flowers you are growing.

- Some vegetables will require more water if there are little rainfall and high heat. Cucumbers, for example, need a lot more moisture than potatoes and onions.
- Flowers and vegetables in raised beds can usually be watered with the same frequency as in-ground flowers and vegetables.
- Container plants need water much more frequently. They can only hold so much at a time without becoming too wet, so you should never drench a plant in a pot. Water it until the soil is moist and stops. Unglazed ceramic pots are especially thirsty, as their porous composition soaks up the water—even wicking it away from the soil, where it is most needed.

Weeding is one of those things you must do whether you want to or not. One of the best ways to reduce the amount of weeding you will have to do is to start with ground that is as weed-free as possible. Taking the time to get as many weeds as possible out of the ground where your garden area will be is the best thing you can do for yourself and your plants. Aside from that, by spending a few minutes every other day weeding a few rows in the vegetable garden and a few minutes pulling weeds from your flower beds, the task will never seem overwhelming.

Other ways you can cut down on the amount of weeding necessary include using ground cover under your flower beds and covering the surface with rock or mulch, laying ground cover between the rows of your vegetable garden, filling your beds with flowers, and planting something in nearly every inch of available soil in your vegetable garden. NOTE: The motive behind planting things so close together is to reduce the number of spaces weeds can grow.

Some weddings will have to be done by hand—literally pulling the weeds from the ground with your hands. In larger spaces, you can hoe up the dirt all around the plants and then pick the loose weeds out and discard them.

Deadheading is essential for a pretty flower bed. You should get rid of wilted, dried-up blooms daily. This will allow more nutrients from the plant to be used to make more blooms and healthier foliage. To deadhead, simply pull the dead flowers off the plant with your fingers or snip them off with a pair of garden scissors.

Spraying and fertilizing is another 'must.' Spraying for pests can be done periodically just for safety's sake but is usually only done at the first sign of trouble. By keeping this aspect of garden maintenance to a minimum, you will be allowing nature to do its thing, i.e., letting the good bugs take care of the bad ones. There are exceptions, however, to this rule.

For those of you who are anti-chemical, you can use gentler concoctions such as the dish soap spray I've mentioned a few times. Cayenne pepper and water are effective in killing ants, orange oil mixed with water and dish soap kills slugs and roaches, and tobacco and water sprayed on plants kill aphids.

Fertilizing is something you can do for the soil and your plants. You should fertilize the soil every fall and spring by adding compost, manure, or commercial fertilizers to it. Fertilizing plants should be done carefully—not too often and not too heavily. Too much fertilizer 'burns' plants, meaning it is too strong. Over-fertilizing can weaken and even kill a plant.

If your soil is well-fertilized, you won't need to fertilize your plants—especially if they are in the ground or raised beds. Plants raised in pots, straw bales, or hydroponic gardens will need fertilizer. Liquid fertilizers or fertilizer sticks you bury in the soil are your best options. Just be sure to follow the directions on the package. Or, if you prefer to go a more natural route, you can do any of the following:

- Water your plants with fish emulsion mixed with water.
- Spread coffee grounds and finely crushed eggshells on the soil and mix them into the surface of the soil.
- Bury chopped banana peels in your potted plants.
- Water your plants with pure molasses water (1 tablespoon to a gallon of water), unsweet tea, or Epsom salts water (1 tablespoon to a gallon of water).

Caring for your garden is not all that difficult or time-consuming. Unless you have large areas to care for, the average person can do all that needs to be done in the same amount of time it takes to watch one or two television shows. And it's better to tend to potatoes than to be a lazy person.

List of Common Garden Weeds:

- Dandelion
- Purslane
- Plantain
- Quack Grass
- Crab Grass
- Saw Thistle
- Carpet Weed
- Chicken Weed

Keeping Pests Away from The Container Garden

You have started a container garden; your fruits and vegetables are thriving and along comes a pest. It may be the four-legged version or the slimy, creepy crawlies. There are ways to protect and keep your fruits and vegetables from being destroyed.

Now most gardeners have opted to be chemical-free, but if not, there are plenty of pesticides and insecticides to keep your plants safe. If you like chemical-free, there are plenty of options to prevent those pests from eating your produce and demolishing your fruits and vegetables, which you have spent time and money cultivating.

As a gardener beginner, you will want to look at your plants and examine what is eating them. Remember that ninety percent of insects are beneficial, and we do not want to kill these as they eat the ones who destroy our plants. When using pesticides, keep in mind that all insects both the good and the bad will be killed, throwing off the garden's ecosystem.

The easiest pests to get rid of are snails and slugs. They eat the leaves and leave nice holes in your produce.

The cheapest route is to take an orange and or grapefruit half that is used and place grape juice or beer into them. These slimy creatures can't resist these smells and taste and will delve in.

In the morning, you can pick up the fruit half and snails/ slugs and dispose in the trash or kill your pest and throw it in the compost heap.

If you do not have fruit halves, fill a saucer with the beer or grape juice, mixed with a teaspoon of salt, and these lovelies will dive right in and drown. If this is too much for those of you who love critters, it is said that copper wire or copper taping placed around your pots will keep these creatures at bay.

After some, there is some maintenance required for this to work. The copper must stay clean, and no grass or leaves, or branches can rest or grow up against the pots as the snails/slugs will be able to crawl up into your containers. Then there is the old-fashioned way of just picking them off and moving them to another location.

The one good thing about growing your fruits and vegetables in containers is that you can inspect them daily. Hence you can remove any tomato worms, caterpillars, or any other destructive insect that you may see, leaving behind the ladybugs and praying mantis to do their job.

Organic sprays have also come along way. Several good varieties are Rose Farm, peppermint spray, which smells wonderful but keeps insects at bay, 100% garlic oil, which is sprayed on the leaves that are being chewed, preventing further damage.

There are insect soaps that are sprayed on the bugs, making them vulnerable to their surroundings. If spraying, picking, and examining your plants is too much work, there is a row cover that is lightweight that can be placed over your plants, allowing air, light, and water to come through but protects your plants from pests.

Unfortunately, this takes away from the aesthetics of garden containers, but they could be used at night, allowing your plants to be protected from evening pests.

The most common complaint in gardening is squirrels and chipmunks; these cute critters will demolish flower bulbs in minutes, eat your greens, and just ruin your mood for gardening. The easiest route is to place wire mesh over your containers before they have sprouted up.

Make sure to secure firmly as these pests are extremely intelligent. River rock and larger gravel also seem to keep them out of containers, and it gives your containers a very natural and pleasing appeal. Like other animals, squirrels and chipmunks do not like strong odors.

Coffee grounds are a great deterrent and are found in almost everyone's home. The grounds can be mixed in with your soil, or spread around the parameters of your container garden. Coffee is also high in acidity, so it is extremely good for most of your garden plants. Soap bars are another great way to keep four-legged pests away. Tied on a string or laying around containers will deter.

Try another smelly deterrent that will make rabbits, skunks, squirrels, and deer stay away. Fox and coyote urine will deter most mammals as they fear that a predator is lurking in the area.

This spray can last you all season and keep the pest at large. Reflective tape, anything that moves or makes noise will also help with annoying animals and send them scurrying into your neighbor's yard or up the street.

Another easy solution is to place a birdbath near your containers, the bath will attract birds and keep bugs at bay. This lovely addition will also give you complete happiness as you get to see a variety of birds come around.

Plant flowers that will draw in the good bugs, such as bees and butterflies, to pollinate your containers, along with the flowers. Plant herbs that will distract the bad bugs from eating your treasures.

Conclusion

Thank you for making it to the end!

Having read this book, hopefully, you have gained a much greater insight into all aspects of fruit gardening. For those who truly enjoy the hobby, it is very much an art form and is something that is relaxing and inspiring. Hobby gardening is practiced across the world by people of many different cultures, and fruit gardening has been a great way to open that practice up to people in urban environments, renters, and people who reside in areas with no fertile soil.

Hobby fruit gardening is also remarkably popular with children, the elderly, and the disabled, as it gives people a chance to explore their creative sides and to nurture and care for another living thing, and to watch it thrive.

Fruit gardening is a great way to show off your artistic skills. Many people find themselves with a few creative outlets when trying to juggle a career and a family.

For many of these people, the simple act of growing plants and caring for them can be a great way to release frustration and express their natural creative abilities. Creating elaborate arrangements and putting them out for the entire world to see can be remarkably beneficial.

In addition to benefits such as expression and edible plants, there are many health benefits to fruit gardening as well. The act of caring for and tending plants can go a long way towards fighting anxiety and depression for many people. It encourages the mind and body to slow down and to appreciate the small things that surround us. It is excellent for people of all ages and can have a significant impact on both physical and emotional health.

When it comes to kids and teenagers, gardening is great for teaching lessons about earning rewards. When a young person is taught through experience that by giving to these plants every day, they can be rewarded with fresh fruits and vegetables that can provide them and their community with nutrition. Gardening can also help them to foster a healthy respect for nature and the environment as well as a healthy attitude towards learning.

For elderly individuals, gardening offers many physical and emotional benefits. In addition to benefits such as lowering depression rates, gardening helps to eliminate boredom and loneliness by providing a rewarding activity that can be easy to do. Physical benefits of fruit gardening include improved dexterity and hand strength, as well as improved coordination and eye strength. For people who exhibit low physical activity, the exercise provided by moderate gardening can be enough to help improve or maintain health. Many health institutions have recognized these benefits, and gardening programs are becoming more common in senior living communities as well as gaining more recommendations from general physicians.

Another benefit of fruit gardening is that it can be an excellent source of food or herbs. Many people grow vegetables, fruits, and herbs in containers both inside and outside of their homes. When you create a successful garden, you will likely be surprised to find how few plants it takes to provide your family with the vegetables, herbs, and fruits needed to sustain you through the growing season. Many people choose to cultivate extra plants so that they can or jar fruits and vegetables or dry herbs for use during winter.

Still, others choose to use fruit gardening to brighten and beautify their homes and property. There are endless ways to use plants to freshen a space and to add color and flair. Whether lining a driveway, creating a flower bed or covering the foundation of your home, there is much you can do to use fruit gardens to enhance the outside of your home. Many people create elaborate landscaping designs using only container plants. With an eye for color, an understanding of what plants will thrive best in your area, and a knack for picking the right containers for each plant and space, you can certainly achieve a yard or area that looks as though it was designed by professionals while spending only a modest amount of time and money.

The benefits you see with your fruit garden will mostly be determined by your expectations. If you step into this project or hobby, expecting to put your plants outside and watch as they continue to flower and bloom, you are likely to experience disappointment. If, on the other hand, you expect to invest time and effort feeding, watering, pruning, and caring for your plants, you are apt to find yourself quite satisfied. Plants are much like people in that they must be nurtured to thrive.

No matter what your age, there is much to be gained through gardening. This hobby offers an array of physical and emotional health benefits and can provide a great way to boost self-esteem while lowering rates of depression and suicide. Caring for plants can be a great way to give back to the world around you while enjoying an activity that is relaxing and cathartic. Whether you are seeking a way to alleviate stress or a chance to enjoy the fresh air before work, tending container plants offers an ideal solution.

In all, the benefits of fruit gardening are simply immeasurable. Whether you are seeking to implement a few hanging garden baskets in your home or on your porch or are seeking to create a beautifully landscaped front yard, you will find that gardening is simply good for you.

Learning to master fruit gardening is something that takes a lifetime but learning to enjoy and reap the benefits can be done quickly.

Indoor fruit gardens can be just as rewarding. Whether you are experimenting with organic fruit gardening to create fresh herbs or simply beautifying the rooms in your home, you will find that there are many choices available inside your home. Window boxes and displays can make every room brighter, and a simple plant hanging in a hallway or bathroom can add a splash of life to notoriously drab spaces. With a bit of flair and the right plants, you can draw the eye to the parts of your home that you feel go unnoticed, ensuring that every room of your home is inviting and welcoming to all who enter.

No matter your reason for creating a fruit garden, you will find the hobby quite rewarding, provided, you can invest the time and care needed to develop healthy and thriving plants. This book was designed to help make creating a fruit garden more fun and more rewarding. You will quickly learn that fruit gardening is a great way to get fast and easy results. While maintaining plants requires a bit of work, the effect of placing beautiful arrangements in your home or yard is immediate.

Unlike cultivating an outdoor garden, you will also find that fruit gardening can be perfect for people of all skill levels.

Even if you have never been able to keep a standard houseplant thriving, following the tips and guidelines you have learned within these pages can help you develop a green thumb. Choosing plants that have low care requirements is always great for beginners, and in no time, you will be able to grow beautiful bulbs and flowers with the greatest of ease.

As with any hobby or undertaking, one of the secrets of fruit gardening is to be patient with yourself and your garden. No matter how much effort you put forth, plants will bloom only when in season. Take the time to appreciate new growth and celebrate small accomplishments. When you can care for your plants throughout the year and take the time to prune and deadhead them as needed, you will certainly be rewarded with a bright and long blooming season.

It is important to remember that mistakes and setbacks can happen as well. If you overwater or over-fertilize a plant, take the experience and learn from it. You will not always be a perfect gardener but understanding that even the most advanced gardeners make mistakes will help you move beyond these situations. The needs of your plants can change with shifts in weather, temperature, and climate, and only with experience can you learn how to anticipate what changes you will need to make in caring for them.

You will also need to remember that some seasons will simply be better than others in terms of outdoor fruit gardens. When there has been an extremely hot and dry summer, you can expect to have a harder time maintaining your plants. In many areas, this type of season can spell certain death for all but the hardiest of plants. Don't let rainy or hot seasons and the changes they bring discourage you from gardening. Nature is always going to be difficult to predict, but the benefits of gardening are quite rewarding. If your plants do not survive a particularly hard winter, simply create a new garden and start over. Using more insulation or learning what you can do to prevent the problem is always much better than simply giving up.

As climates change and new plants grow in popularity, there is always something new to be learned in the art of gardening, ensuring that even the professionals will always have more to understand. With the tips and tricks, you have learned in these pages, however, you are certainly well equipped to begin a fruit garden of your own. As with any hobby, the most important things to keep in mind are simple to be creative, take your time, and have fun.

CPSIA information can be obtained
at www.ICGtesting.com
Printed in the USA
BVHW082327030521
606340BV00008B/2131